make it fail. Relativity actually does fail in its predictions- it is a theory which cannot survive unless we live in the world of dreams and fantasy- but even then it would be a really silly world fantasies- devoid of logic and common sense- which children can enjoy laughing at!

To overcome this clear contradiction and to cover up for Relativity, Engineers and scientists at CERN propose a multiplicative factor (which equals zero) that we "have" to multiply to any velocity that might lead to velocities more than light. Yes indeed, Special relativity is a dead theory- now made artificially alive by ever loyal scientists to fake beliefs!

We all know that a theory requiring constant "UNJUSTIFIABLE" corrections is a failing theory; we must not keep adhering to. This reminds me of the cosmological constant that Einstein proposed by himself to justify his inability to believe that the universe is expanding, just to assert later on that "it was" the biggest "blunder" of his life, of course that was after Edwin Hubble discovered that the Universe is actually expanding and Galaxies (with the exception of close galaxies affected by each others' gravity) are all receding from each other at ever increasing speeds making our observable universe an expanding one.

We must expect from Einstein to have said the same – regarding the CERN outcomes- that is if he was still living. But he's not here anymore, we are left with mostly obedient scientists, who hate change, and who don't aim for the truth nor for advance of science. Most of these concerned scientists are just afraid of facing the ridicule if they openly admit the inconsistency and the contradiction Relativity brings- a very painful fact indeed.

E. ACTION AND REACTION OR ACTION ONLY

The following is a section I liked to include- it is a question I enjoyed thinking about since long. It is believed that Rockets actually propagate in space according to the action and reaction principle, which simply states that for every action there is an equal and opposite reaction. While this principle surely holds in the world around us, it might actually not hold for rockets.

Explained simply, when rocket fuels burn, gases are ejected from the nozzles of the rockets fired upwards, and if we consider the system of (Fuel- Rocket), the gases generated from burnt fuel apply pressure in all directions- against the walls of their containers(internal body of the rocket), just like gas in a balloon does, but surely these gases don't apply their pressure against nothingness or the "opening" of the nozzle, rather, the atmosphere will be receiving this pressure, and that creates a compaction for the layer of air underneath.

So what actually propels the rocket upwards? It is simply

the pressure resultant on the internal walls of the rocket generating a net zero horizontal force, and a net upward given force since the pressure downwards is applied against the atmosphere (as shown in the diagram next page). Rockets therefore are propelled by the action of burnt gases on the upper walls of the containers and not by the "imaginary reaction" that does not have a reason to exist.

The travel of rockets in free outside space- where matter is so scarce, and where space is practically vacuum provides a good reason for me to believe that the rockets are driven forward by the pressure the fuel generates upon combustion and not by the imaginary reaction that vacuum simply cannot apply on rockets.

TAREK SAMI AHMADIEH

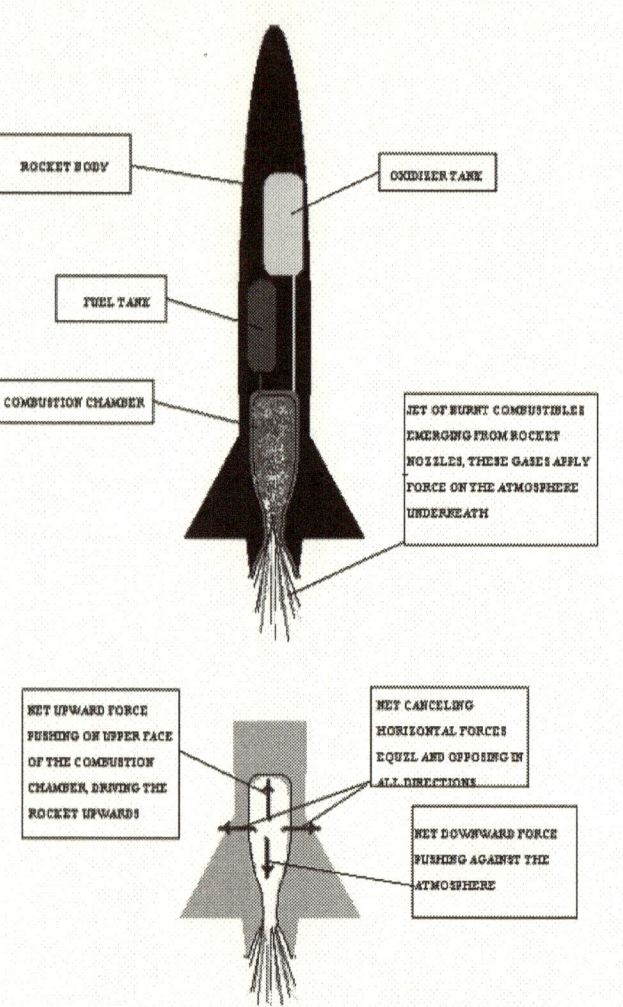

ROCKET PROPULSION THROUGH NET UPWARD FORCES INSIDE THE COMBUSTION CHAMBER

ROCKET BODY

OXIDIZER TANK

FUEL TANK

COMBUSTION CHAMBER

JET OF BURNT COMBUSTIBLES EMERGING FROM ROCKET NOZZLES, THESE GASES APPLY FORCE ON THE ATMOSPHERE UNDERNEATH

NET UPWARD FORCE PUSHING ON UPPER FACE OF THE COMBUSTION CHAMBER, DRIVING THE ROCKET UPWARDS

NET CANCELING HORIZONTAL FORCES EQUAL AND OPPOSING IN ALL DIRECTIONS

NET DOWNWARD FORCE PUSHING AGAINST THE ATMOSPHERE

3. RELATIVITY VERSUS REALITY

.. and we must climb the hill to see the other side....
Tarek S.Ahmadieh

A brief history of Relativity:

1. Relativity is the alternative to common sense
2. Relativity had solved paradoxes
3. Relativity is a simple explanation to the absurdities
4. A renowned scientist had said that he and Einstein were the only people who understood Relativity in the early 20th century.
5. Any observer has the right to claim that he is stationary in his frame of reference, and that the others are moving.
6. Light speed is indeed invariable to all observers- and will always be measured the same.
7. Light speed is the speed limit matter and energy can travel at
8. Light, unlike other electromagnetic waves does not need a medium
9. Light cannot be seen standing still or an observer cannot move next to a light ray
10. Light has a particle nature

11. Gravity is a distortion in space
12. Gravity, like everything else cannot travel faster than light.
13. Time and space are intermingled together
14. Space is described by Reymann Geometry, Eucledian Geometry fails to describe the space we live in
15. Motion back in time is possible, since time is a sensation, and is by itself relative to the observer
16. Simultaneous events must be seen simultaneous by observers
17. Any star like our sun Bends light rays due to gravity
18. No addition of speeds will amount to more than C- the luminal speed
19. No universal reference frame will ever be proven to exist
20. Masses of speeding bodies increase relativistically
21. Lengths of speeding objects shrink in the direction of high speed
22. Time slows down with respect to the fast moving observer- as seen by a stationary observer
23. Atomic clocks are the correct measure of the passage of time.
24. Mechanical clocks will slow down at high speeds
25. Light speed is independent of the speed of its source
26. $E = mc2$
27. Relativity describes the real world

THE FALL OF RELATIVITY

-Whereas-

A brief history of reality goes like this:

1. Relativity has nothing to do with common sense, and is the alternative word of non-sense

2. Relativity had created paradoxes whole encyclopedias cannot accommodate!

3. Relativity eventually creates absurdities from anything easily explainable!

4. Thanks God there wasn't more than two who digested the craziness of relativity back then. If there were, It would have been terrible if more had become relativists, since Relativity has shown to spread among the scientific medium as well as among the general public like bird flew nowadays...

5. No observer can claim he is stationary, which could be valid only for calculation purposes- while it is physically untrue, since any observer can actually measure his speed relative to the universal fixed frame, as shown by recent quantum mechanics experiments. I want to assure that if we feel nothing strange when we are riding on Earth, and if physical laws are the same as for a stationary observer, that does not give us the right to claim we are stationary in our reference.

6. Light speed relative to observers is never the same to all observers, even if it is measured the same (in that case the

experiments are not yet perfected to detect the differences in light speeds)

7. Super-luminal speed had been shown and verified

8. Light does need a medium like all other electro magnetic waves or else it its speed will not be independent on the source of the light.

9. There is no problem at all with a traveler moving at the speed of light, just as there is no problem at all with bodies moving at the same speed of their sound waves, or even breaking the sound barrier. At any rate the light will not appear to stand still as Einstein once thought- because light rays are actually invisible unless they point from a luminous source at our retinas or reflect from matter bodies into our retinas.

10. Light, while it is believed to have dual properties of particles and waves, cannot be considered as particles since that would mean that the light speed depends on the state of motion of the source and in that case Relativity falls altogether- like a house of cards- as Einstein used to see in his nightmares. The famous train and embankment experiment of Einstein will then be explainable in a manner that removes the alleged paradox of relativity that Einstein claims the solution through the awkward use of Lorentz transformations fitted with the invariable speed c with respect to the observers. So if light is particles, Relativity fails, and when light is waves, Relativity has no reason to be,

both way failure, what a theory!

11. Gravity does not warp space, it acts directly on other matter bodies, and it is a feature that may be traveling at faster than light.

12. Gravity may travel faster than light.

13. Time and space are different entities of different nature, space is tangible having properties, whereas time is a smooth uniform flow of events into the future, fully abiding to the cause and effect principle, time cannot stretch nor turn back as Relativity predicts.

14. Reymann's geometry was a cover up for General Relativity's failure to be described by Euclidean Geometry.

15. Back ward motion in time occurs only in the minds of the insane, that is the most serious threat to our well being of our logic and common sense, nobody can go back in time to shoot the one who killed him.

16. There is no problem with simultaneous events being felt or seen at different times; in the same way there is no problem in observing now how the stars actually were billions of years ago.

17. Gravity is not proven to bend light, the cause of light ray bending by our Sun was refraction, and not gravity as Relativity predicted.

18. Velocities may be added freely, relative velocities may add without relativistic corrections.

19. Universal fixed frames of reference are already proven

to exist by modern quantum physics experiments, and have been anticipated by the non- null results of the MMX repetitions.

20. Masses of speeding bodies do not increase, this is only twisted and wrong explanations of experiments dealing with particles. A grain of rice will never have the mass of a Sun like ours if it moves at speeds close to speed of light, that is non-sense.

21. Measured lengths of moving objects do not shrink and neither do the lengths of these moving objects.

22. Relativity gives the right for the any of the two observers to claim himself stationary, and each to see the other shrinking, and getting slower, the two observers will actually kill each other after they argue the paradox Einstein had thrown them into, each of them swears that his time is faster, and each will accuse the other of being of the same size as his son. That is a stupid paradox, digested only by faithful relativists.

23. Atomic clocks are the best clocks humans have till now; these clocks may be affected by high speed motion, having to do nothing with real absolute time passage.

24. There is no reason mechanical clocks will move slower at high speeds, I have read tens of books on relativity but I have seen no convincing evidence of that.

25. While this is true, and when Relativity is based on this fact, it will bring its own down fall. Light is no longer particles

as relativity claims, and being so, it will be waves which require a medium, which is the fixed frame of reference, and then Relativity fails because it does not admit this reference frame.

26. E=mc2 is a formula based on c being an upper limit of speeds, which has been proven to be a wrong supposition to start with.

27. Relativity is actually non sense!

This chart shows the flow of reasoning initializing and following the famous Michelson-Morley Experiment (the MMX):

LIGHT SPEED IS INDEPENDANT OF SOURCE -PROVEN- DESITTER EXP

LIGHT IS WAVES, NOT PARTICLES

A UNIVERSAL MEDIUM IS NEEDED

TESTING FOR THE MEDUIM-THE MMX

1.NEGATIVE RESULTS AND POSSIBLE ALTERNATIVE EXPLANTIONS

UNSUITABILITY OF THE MMX SETUP TO DETECT A MEDIUM- SEVERAL REASONS

INCONCLUSIVE MMX

A UNIVERSAL MEDIUM IS NEITHER PROVEN NOR DIS-PROVEN

ONE BASIS FOR RELATIVITY DOES NOT HOLD

SPECIAL AND GENERAL RELATIVITY COLLAPSE LIKE A HOUSE OF CARDS

2.POSITIVE RESULTS OF MMX

EXISTANCE OF A UNIVERSAL MEDIUM

LIGHT SPEED VARIANCE AND DEPENDANCE ON STATE OF MOTION

SPECIAL AND GENERAL RELATIVITY COLLAPSE LIKE A HOUSE OF CARDS

3.NEGATIVE RESULTS FOLLOWED BY BIASED EXPLANTIONS

NO PREFERED FRAME OF REFERENCE-SPECIAL RELATIVITY

LIGHT IS PARTICLES

DEPENDANCY ON SPEED OF SOURCE

SPECIAL AND GENERAL RELATIVITY COLLAPSE LIKE A HOUSE OF CARDS

LIGHT WAVES DO NOT NEED A MEDIUM

LIGHT WAVES CAN ADVANCE IN VACUUM

CONTRADICTION TO MAXWELL'S THEORY OF LIGHT

WAVES CAN ADVANCE AT THE SAME SPEED OF THE SOURCE – CERN EXP(CHAPTER 2)

OBSERVERS AND SOURCES CAN MEASURE DIFFERENT LIGHT SPEEDS

SPECIAL AND GENERAL RELATIVITY COLLAPSE LIKE A HOUSE OF CARDS

4. A BRIEF HISTORY OF LIGHT

... And should light forever be our limit??? If so- we might forever stay in the darkness...

Constant, unchanging, in-exceed able, it is the free speed of travel of light, referred to as c, It is just a number nature has chosen, just like Pi, the ratio of the circumference of a circle to its diameter, and nature wanted the value of c to be around 186,000 mph (Even though there are reports of higher comparative light speeds- hence directly contradicting the Special theory of Relativity).

Danish astronomer Christensen Roemer had showed in 1676 that the speed of light was finite and not infinite; but not until 1865 was James Clerk Maxwell able to provide the theory of the electro magnetic nature of light traveling in waves; Maxwell showed that light propagated in waves as disturbances in an electromagnetic field. Maxwell's theory was consistent with the finding of Keneth Brecher at Massachusetts Institute of Technology in 1977, showing that light is unaffected by the speed of its source, and that reinforced the fact that light had wave properties. Brecher's finding was through observing X-Ray pulsars flashing regularly and periodically through space. But in 1905, there came Albert Einstein to suggest that light speed is unreachable no matter how fast an observer is moving, and no matter at what speed the observer is moving light always overtakes him at the speed c, no more and no less!!!! That was the core idea of special relativity.

Going through a medium, light's speed c can be expressed as a function of the magnetic permeability and the dielectric constant of the medium which light travels in, that is just analogous to the velocity of transmission of vibrations through solid bodies where the speed of propagation of sound waves are function of the density and the elasticity of the substance the waves are moving in-(Appendix-3).

H.E.Retic explains the interesting paradox of constancy of light measurement through showing the actual cause of in-variability of light speed measurement, he shows that any attempt to measure the velocity of light C will ultimately yield the same result because all experiments done till yet can only enable us to measure the fine structure constant (the expression of which is shown in appendix-3)! Light speed, hence, is related to the fine structure constant, the planks constant, the charge of the electron, and the dielectric constant- which all matter properties permitting matter to re-arrange itself to satisfy this relation that will lead to a C=C type of measurement! A balance showing your weight to be equal to some other entity's weight without telling you the actual value of the weight is actually is useless balance; similar to the light speed measurement experiments being performed. So in effect, there is a cause for the illusion of the constancy of light speed – an explanation for the "null" result of the MMX! Our experiments are not yet fit to do the job, we can only wish that we might be able to measure the exact "actual" speed of light relative to us (the observer) in the near future. We have to keep in mind that the absolute velocity of light through space is inevitably shown to be around 300,000 km/sec.

For me personally, believing that "c" is un-reachable was hard, believing that it was in-exceed able was tormenting, but admitting that no addition of velocities ever can surmount "c" was something at which I never felt at ease with. Apart from my basic intuitive feeling that there must be serious misinterpretations that lead to this belief, and apart from the fact that this didn't blend with logic; I felt sure that the reason is beliefs which we became reluctant to question. Hence, many things are becoming far from where it ought to be.

As I mentioned previously, Einstein wondered what would happen when an observer A moves at the speed of light, adjacent to a beam of light, and in the same direction, would this observer see light standing still besides him?!? Here *there is another important question that jumps into the picture: "Does light really overtake any moving observer at the same speed as if the observer is standing still?"- as Einstein believed.*

A. SEEING LIGHT - WHICH BY ITSELF CANNOT BE SEEN!

In order to answer the questions posed in the previous paragraphs, I want to ask this apparently very trivial question: *When do we see light? I am not really sure how has this question been asked before- if it was ever considered, but in my opinion:*

We "visually sense", or "see" light in the following 2 cases:

1- Seeing light rays being reflected from bodies- non luminous matter- no matter how large or small it is, (and of course we mean by light the visible waves of the electro magnetic spectrum)

2- Looking at the light source – luminous matter itself. On the other hand, we cannot see light rays moving freely through space and not heading towards our eyes and not striking our retina, it simply cannot be... I was stunned when I read in one book popularizing relativity by the power relativists posses when it comes to twisting our minds and convince us by using arguments as invalid as the theories they try to prove. Let us check out one of these arguments, and notice the unjustifiable jump from one conclusion to another: He (Einstein) inferred another curious effect concerning the speed of light. When the speeds of objects approach the speed of light, you cannot add them together in the obvious way. Picture two galaxies rushing away from the earth at seventy five percent of the speed of light each- in opposite directions, away from our Earth, simply adding their velocities will give us that the galaxies are actually recessing from each other at one hundred and fifty percent of the speed of light. In that case, you might think that one galaxy must be invisible from the other because light passing between them could never catch up.

But...the argument continues ... it is easy to see that they are still in contact, because one of them could send a message to the other if need be by the way of the earth. So in effect, the relative speed of the galaxies must be less than the speed of light, because messages can still be commuted between the galaxies.

We can easily figure out the flaw in this argument; as the speed of light is well known to be un-affected by the speed of the source, so no matter what the speed of the galaxy the light starts from, this light will travel at the c speed, it will sure reach the other galaxy darting in the other direction after some time, and when the other galaxy is moving at seventy five percent of the speed of light, the light will reach it and by pass it at twenty five percent of the speed of light .... So the two galaxies can communicate by light waves, but does this communicatability mean that the relative speeds of the galaxies are not algebraically additive?!?!? Surely not, the relative speeds of these two galaxies is one hundred and fifty percent of the light speed, nevertheless, light starting from one of them towards the other will catch up with the other galaxy and the reason is that the light doesn't "feel" the motion of the source galaxy-thanks to the wave properties of light and electromagnetic radiation in general. So relativists erroneously interpret this virtual experiment, and unjustifiably try to use it for proving that high speeds need not be algebraically additive - and that is their excuse to apply the superficial contraction factor of Lorentz to the distances, and to the timing and masses of moving objects!

Relativists suppose that light starting from the source galaxy will by pass the other galaxy at the speed of light no matter what the speed of the other galaxy is!!!! They explain that this is manifested by the fact that an observer on the receiving galaxy will measure the speed of the beam of light from the source galaxy and find out that it is equal to c! Their best reason is the wrong explanation of the Michelson Morley Experiment - the MMX...

But was the experimental setup of the MMX fit and suitable for deciding on the invariance of the speed of light?!?!? Surely not, especially that later repetitions of the MMX experiment showed positive results that make relativity fall like a house of cards- that is exactly what Einstein feared himself!

So, can we suppose that a galaxy can go faster than light with respect to absolute rest or to any other entity in the universe? Relativists say no!!! Their reasoning is that time will be running backwards!!! What reasoning is that?!?!?! Who said that time slows down for fast objects in the first place??? Even kids know that when they have to prove something, they must refer to something accepted and undisputable, and not to something questionable and debatable.... It seems that relativists must use "their" invalid relativistic assumptions to prove their other invalid points of view- they lack basic understanding of physical laws, logic, and experimental verification. Relativists do breach logic of argumentation in every argument they conduct.

B. AN OBSERVER MOVING AT THE SPEED OF LIGHT

Hence, observer A, when moving at the speed of light in parallel to and adjacent to the beam of light, would have seen nothing unless when this beam struck matter bodies or fragments of matter on its way. If the observer moved his arm perpendicularly to interrupt the light rays moving next to him since his departure together with it, he would see his arm by reflections of these same light rays.

Theoretically, our logical reply to Albert Einstein will be: Observer A could have not seen any standing light- for light is moving after all, but along side this fast observer, and <u>NOT OVERTAKING HIM</u>. Light doesn't stand still at all; but light surely wasn't overtaking observer A at 186,000 mph. In striking similarity to the "recent" proof of Gamma rays moving along side the pi-mesons in the CERN laboratories (explained before), the light rays will have moved next to the observer A-with no single reason to doubt and wonder how that could be-that simply is something very logical and this phenomenon is no affront to any scientific principle concerning behavior of electromagnetic waves.

For people who do not have an idea about special relativity, the above argument is perfectly sound and rational. I will move on to answer those who already know what special relativity claim: In his theory, Albert Einstein claims that light is able to surpass any moving observer, no matter what his or her speed is, by its own speed. Light, according to Einstein would move past a moving observer by its own speed, and the observer would see the beam of light overtaking him at the speed of C, the speed of light.... In Einstein's world, observers will always measure the same speed of light no matter what the direction and speed they move in!!! So Einstein's light rays guess at who is observing them, and accelerate and decelerate accordingly to satisfy the constancy of measurements depending on who is measuring its speed!!! And what a conspiracy that is against science, logic, I think that relates to hallucination.

More than that, Einstein's special relativity stipulates that no addition of relative velocities might exceed the speed of light........ Had we only known that this number "c" number *should* not be surmounted!!!!!

As for the addition of velocities, we could easily prove that speed of a beam of light relative to another beam is additive and when added amount to twice "c"! Let us consider the Earth at the Perihelion (the point on its orbit when it's closest to the sun, and the other opposite point- the Aphelion (point on the orbit farthest from the sun). At a specific moment t, light is ejected from the sun in both directions A and B, beam A will ultimately reach the perihelion some seconds before beam B reaches the aphelion on the other side. At the moment t1 at which beam A reaches the Perihelion, the light beam B would have advanced at a distance = perihelion distance – the same as the distance light has spread in all directions – concentrically from the sun. Light beams A and B would be = 2 x perihelion distance, and having advanced at the "c" speed in both of the opposing directions, eventually the two beams of light had moved at twice the speed of light comparatively (Vb/a (Velocity of beam b relative to velocity of beam a)= Va/ b (Velocity of beam a relative to velocity of beam b)=c+c= 2c). That is an example of light relative speeds higher than c- a clear contradiction to Special relativity. A simpler example would be light from any star X approaching light coming from another star Y, rays of light pass each other at the speed of twice that of light- very clear, no further explanation. I want to assert that even if we take the alleged time slow down of Relativity into consideration and if we suppose that time slows down for the particles of light, and we will also assume that the measuring yard stick with respect to the moving particles shrinks in the direction of motion- that is according to special relativity)-that would simply mean that light would be traveling longer distances in less time translating to an even greater speed than c- yielding even more contradiction to Relativity itself- Relativity explodes by itself again- what a special theory it is- very special indeed!

That is true, Albert Einstein had un-intentionally confined us between the boundaries of light, he had trapped us there, and he forced us towards this perfectly illogical paradoxical world... Einstein sure did not plan that, he *was sure searching for the ultimate truth, he suffered for that. Fortunately for him, and unfortunately for 20th century science, Einstein reached the conclusions that lead him to put forth his special and general relativity theories.* But how did Albert Einstein conclude that neither velocity nor any addition of two velocities can exceed the speed of light? What lead him to think that way? What I and you wish now is that he was still alive so we could have asked him..... Of course, that would have been a lot better...

We can, alternatively read his book titled "Relativity- the special and the general theory", we can quote from his book: "Can we conceive of a relation between place and time of the individual events relative to both reference bodies, such that every ray of light possesses the velocity of transmission c relative to the embankment and relative to the train"- in his famous observer and train theoretical experiment. Einstein then leads us to his generalizations using the above quoted seriously questionable assumption (Einstein supposes that Light speed should always be measured the same, by any observer- because all observers should be given equal chance of knowing the truth about any physical measurable quantity especially speeds of other moving objects, the alleged "null" result of the MMX had eventually encouraged him to do that)! That is purely twisted logic or no logic , I personally see that any observer has his or her privacy of measurement, and his answers for measuring speeds of other moving objects should be affected by his state of motion, that is what Relative speed measurement is all about). So Einstein's Relativity is not Relativity after all – it is the non-variance theory as he used to call it himself, and in any case is wrong! There should be variance, Einstein, after all had not understood exactly what relativity should be talking about, it is about variance. We have to make sure we do not mix the relative measurement of quantities with the laws of physics being the same and holding true for observers. Physical laws hold every where even when different observers measure different entities and get different answers, as long as observers do not affect physical properties by merely watching the entity itself!!!

I want to clarify here that if, for the moment, we take the (v+c) speed of light into consideration – meaning light reserving or "feeling" the speed of its source (that is confirmed by some Radar measurements- brought up in the section about superluminal light speed measurements, the alleged paradox that bewildered Einstein himself, and led him to generate the invariance theory will be solved. Now since Einstein (and we) nowadays still want to stick to the notion of the independence of the light speed from the speed of the source, we can just say that the two observers sitting on a moving carriage on a rail wail with a lamp in between, at midway in between, will simply not see the lamp light simultaneously, that is because the motion favors light rays going in one direction, but not the other, that is why one observer sees the lamp lighting before the other, there must have been no paradox for Einstein in that. No need for the awkward invariance of light explanation, and Einstein has no justification to wonder into fantasy just because he had the feel that the observers "should" see simultaneous events at the same time, that is just non-sense.

Einstein, in effect uses Lorentz's transformations to show that time slows down and distances shrink in order to satisfy his constant speed of light – for "c" must not vary, that is the essence of Special relativity- based on the extremely serious assumption of invariance of luminal speed!!!

C. SIGNALS CAN TRAVEL AT SUPERLUMINAL SPEEDS ...RELATIVITY FAILS EASILY!

A story: A genius little green creature makes a rod of negligible weight out of the exquisite material it found on its distant planet in a far away galaxy from our milky way....... The Rod's length, the creature wanted it to be 186,000 miles .... (Sure very long compared to our earthly daily life common things).Holding this rod in his powerful hand, it points out to the distant stars, where it sees a faint light source of spiral shape!

The creature is pointing at our galaxy, and it says to another watching creature.... "I wonder if there is some sort of life on any planet in that galaxy ...", and he shakes the rod outwards towards the sky... Actually this talk is dangerous, not because there might be creatures out there, but because these creatures "play" is a sufficient proof that a humble mechanical movement could be transferred to a distant place at more than light speed... To readers who know and those who do not, in their books explaining relativity, relativists assert that no signal whatsoever can be transmitted at speeds higher than c, however one of the main aims of this book is to focus on the the possibilities super luminal speeds! Assuming that the creature shook its rod in half a second, and assuming that the movement was felt at the other end of this Rod (of course we have to consider that the rod is in compressible, and sufficiently rigid...), then this little mechanical movement or "signal" was transmitted through space at two times the speed of light ! Moreover, the creature shook with its rod, without knowing (or on purpose, and in that case, it's a shame.), one or more of our pillars of modern physics. That signals that we are taking things too much for granted here on this blue planet.

Most of us had seen billiard balls; we imagine these balls stacked one after the other in an array, each ball just touching the ball in front and behind. We can push the first ball, and we can observe the last one move forward by the same magnitude. Now we can imagine an infinite number of balls- or merely an array of balls extending 300,000 kms long, the push on the first ball can be felt by the last at the other end instantaneously! That means that the pulse had been affected at 300,000 kms away almost instantly- in less than a second- and that is of course a signal moving at speeds folds and folds higher than the speed of light. That is a humble mechanical movement transmitted over large distances faster than light- once again. I can propose that gravity effects are transmitted between celestial bodies in the cosmos through much the same way, and felt instantaneously once any two bodies simply exist. We don't have to imagine a cosmos of space-time fabric warped by existence of bodies, or a space that act as a transmitter of gravity like what Einstein supposed when he rejected the fact that gravity actually does move at much higher speeds than light does. After all, the graviton particles – which are by definition the hypothetical particles or the quanta of gravitational energy , predicted by the general theory of relativity, are supposed to act as transmitters of gravity, but have not been proved to exist till yet. These gravitons are predicted to move at the speed of light and to have zero rest mass and charge and pictured as the gravitational equivalent of photons.

I have to mention here that to the best of my knowledge, there is no scientific evidence or measurement revealing the speed of propagation of gravity through space. We do not know how does gravity propagate in space till yet, all we have is theories still very far away from experimental verification.

D. PARTICLES TRAVELING AT SUPERLUMINAL SPEEDS

Turning to our genius's works, in Stephen Hawking's "Brief History of Time", there is a chapter called "Black Holes Ain't so black".. In this chapter, Hawkins assures that certain particles are allowed and actually "do travel" at faster than light for some distance so that it can flee away from the tyranny of the ultimate space carnivore! By doing that, black holes seem to radiate, tossing off particles to space, where the particles return to move at lower than light speed!

E. LIGHT IS BENT BY REFRACTION NOT BY GRAVITY

Einstein was reported as saying that he slept during the eclipse of 1919, he was so sure that the observations will confirm that light rays originating from a distant star and passing tangent to the sun's surface will be bent by its powerful gravity. The results showed indeed that the light rays were really bent by 1.75 arc seconds....So light is indeed heavy and gravity of our Sun had bent its path!!

We all know that light, when passing from one medium to another heavier medium, is refracted from its original path. This important scientific proven phenomenon was completely overridden in the interpretation of the light ray deflection observations of 1919, making one of the biggest blunders of the human scientific endeavor. When we actually consider the refraction due to sun's surface heavier vicinity, we discover that this bending observed was due to refraction and not due to sun's gravity (Unfortunately only a small minority of the scientific community nowadays admit that refraction is the actual cause of light bending in vicinity of stars, while they refer to this observed phenomenon by gravitational lensing).

If we recall that Earth's recorded imposed deflection on inter- stellar light rays passing tangent to it is 1.92 seconds of arc, we have therefore to admit that light is bent more by Earth than by the sun, which is more than 300,000 times heavier than earth. So, if we commit to the gravity being the cause of the bending, we are therefore asserting that Earth has more gravity than our Sun, which in turn is much heavier! *Now, it is up to the reader once again to infer, without any discomfort or any suspicion that the observation of 1919 was misinterpreted. The reason behind the misinterpretation was the un-justifiable prejudice and erroneous pre-premises considered to be un questionable, that lead most of the scientific community of 1919, eager to reinforce Einstein's emerging theories to overlook the real reason for bending of light – REFRACTION, and not GRAVITY !!! And here, the second "classical proof" of General Relativity is dethroned, and Relativity again fails miserably...*

When describing the expedition that set out to prove the light rays bending by Sun's gravity, Stephen Hawking says in A Brief History of Time: "It is ironic therefore that later examination of the photographs taken on that expedition showed that the errors were as great as the effect they were trying to measure. Their measurements had been by shear luck or a case of knowing the result they wanted to get, not an uncommon occurrence in science. The light deflection however has been accurately confirmed by a number of later observations". Therefore, one classical "proof" of General Relativity does not hold anymore. But we have to keep in mind, at the end, that light has changed direction (light accelerates - like a race car on the curve of an oval race track), when passing tangent to the sun, it is not that "sacred" anymore...

F. LIGHT IS CAPTURED AT EVENT HORIZONS OF BLACK HOLES

Light is believed to hover at Event Horizons of black holes (so called because nothing can escape their tremendous pull of gravity, not even light itself), so Light can be arrested or stopped at one single zone in space...... If light can be stopped, it will probably not come to a complete halt immediately, rather it will decelerate (since light is believed to have both matter and wave properties). This leads us to speculate about different light speeds existing in the world depending on which gravity zone light is passing in!

Luminal speed, therefore, might no more be considered a reference, and no longer can "c" be considered a fixed speed as once thought. In effect, and in light of what I mentioned above, we have the right to suspect that speed light is not an absolute speed .....Light is not sacred after all ..... It can be bent; it can be stopped, it can be surpassed by humble mechanical movements.... I have to re-mind here that relativity dictates that light measures the passage of time for us in atomic clocks .... We have the right to put the results under real scrutiny and questioning, for the sake of science, logic, and mankind advance.

Shown below is a definition of the "effect of gravity on light" from a site popularizing Relativity on the web:

According to General Relativity, the wavelength of light (or any other form of electromagnetic radiation) passing through a gravitational field will be shifted towards redder regions of the spectrum. To understand this gravitational red shift, think of a baseball hit high into the air, slowing as it climbs. Einstein's theory says that as a photon fights its way out of a gravitational field, it loses energy and its color reddens. Gravitational red shifts have been observed in diverse settings.

Earthbound Red shift

In 1960, Robert V. Pound and Glen A. Rebka demonstrated that a beam of very high energy gamma rays was ever so slightly red shifted as it climbed out of Earth's gravity and up an elevator shaft in the Jefferson Tower physics building at Harvard University. The red shift predicted by Einstein's Field Equations for the 74 ft. tall tower was but two parts in a thousand trillion. The gravitational red shift detected came within ten percent of the computed value. Quite a feat!

The reader might refer to Appendix-4 for details of the Pound-Rebka-Snider experiment carried on at the Harvard Tower, and describing the effect of Gravity on the energy of Photons(whereas I prefer to rephrase as "the effect of gravity on the energy of the Light through elongating the wavelength") - with all respects to light duality description proponents.

It is clear therefore that the effect of gravity on light is evident through the "reddening" effect; the light waves are shifted towards the Red end of the spectrum. Gravity cannot have a double effect, it either affects the wave length or it affects the speed of light, and since light is said to puff at the event horizon of black holes of ultra strong gravitational fields, that means one of two:

It's either:

Gravity affects speed of electro magnetic radiation- the light, but not the wavelength of these radiations: In that case, Stephen Hawking's prediction of light arrested at the event horizon of black holes is right and Einstein is wrong in claiming luminal speed as constant, or that:

Gravity elongates the wave length, and "reddens" the light, and hence Stephen Hawking's prediction of light hovering at event horizons is totally wrong- since the effect of the black hole is pronounced through "reddening" of the light and not through slowing the light speed, and Einstein's prediction that the effect of gravity on light is through elongating the distance between wave crests and hence "altering the energy of the light rays, but not the speed" is right.

At any rate, the talk above is consistent with Einstein's reported un-easiness of his General relativity in cases of strong gravitational fields- AND THAT REMAINS UN-TESTED TILL NOW.

To get the picture better, imagine a policeman chasing a criminal, the policeman traps the criminal on the roof of a multiple story building: he shoots him, and the criminal falls off the roof to the street below. The criminal is eventually dead, but that means one of two, it is either that the unfortunate criminal was killed by the Policeman's bullets or upon his crash in the street below. Chances are very remote that he has died because of both causes. In effect, the second "classical" proof of General Relativity holds only if Stephen Hawking is wrong in that regards, General Relativity does not survive the test of Black Holes, if it does exist out there.

G. THE SAGNAC EFFECT

GPS satellites are adjusted by using the famous Sagnac Effect, yielding sharply correct answers without taking into consideration any relativistic effects!!! The Sagnac effect is a solid and very tangible proof of Aether existence that Relativity supporters find no way out to give an explanation than to suppose an external frame reference system (which is exactly what relativity itself refutes!!!), and thus Relativity opponents propose existence of the Aether indirectly to prove the inexistence of Aether!!! Crazy as it might sound, but true!!! The Sagnac Effect proves that when two opposing light beams make a complete revolution of the earth in 2 opposing directions, one of the beams reaches the detector before the other producing an interfering pattern of fringes, clearly proving that Light actually has variable velocities relative to an earth bound observer – in clear contradiction to the dictum of special relativity which predicts that light's speed should not be affected by the rotation of the earth, and should be measured the same to any observer on Earth. Actually some scientists regard the Sagnac effect as a positive result twin to the MMX!

H. LIGHT TRAVELS FASTER THAN LIGHT!

LIGHT SPEED WAS BELIEVED NOT TO BE AFFECTED BY THE SPEED OF ITS SOURCE... BUT RECENT TESTS SHOWED SUPER LUMINAL SPEED MEASUREMENTS!

To be able to dwell into this, first we have to re-call that light speed is said to be un affected by the speed and direction of movement of its source. But, there have been various reports of experiments and observations showing relative light speeds of more than c!!! And, still many of these results are being hidden from public and are subjected to powerful anti-propaganda!!!) . I have to say that the temple of relativity crumbles like a house of cards if these experiments –yielding the (c+v) results (meaning ultra light speed measurements of light – which relativity forbids!!!) are spoken out to the public. We will consider for the moment that light rays emitted from a moving source do not feel the speed and motion of the source - Dutch astronomer De Sitter showed that the velocity of light cannot depend on the velocity and motion of the body emitting the light based on his *observations of double stars.*

*Even though there is documented evidence of the independent nature of the speed of light, but the v+c(ultra luminal light speeds) recent radar measurements might turn every theory connecting to light upside down, and much of the physics, as we know it, will have to be re-written! But, supposing minimal source speeds, we will assume for the moment that no matter what is the motion of the source of light, light rays travel at a constant speed – provided that the medium is unchanged and with no singularity such as a black hole being in the route , or else light will experience bending due to refraction, or will be hovering at the event horizon of the black hole.(*The reader may recall Einstein's first speculations of the observer moving at the speed of light rays, or near that speed, which in part was the spark behind his later spelling out of the (special and the general theories of relativity)...

I. ONLY RELATIVITY CLAIMS "C" AS INEXCEEDABLE, NO EVIDENT SUPPORT

I don't know how we easily digest the idea of objects moving at faster than their sound, objects that advance in front of their produced waves "breaking the sound barrier"; like modern assault war planes and jet fighters, which reach to destroy enemy bunkers, before being heard, and at the same time we are un-comfortable with admitting that some objects might or actually "are" moving at faster than the speed of light, and in the same way might reach us, and go by, well before we see it......I recall the section on the wrong explanations of the CERN experimentation regarding pi-mesons, and which provided clue regarding waves moving adjacent to their particles.

J. SUPERLUMINAL SPEEDS AND DARK MATTER CANDIDATES

We might be seeing light rays zipping in our skies, the sources of which are moving at ultra-light speed motion (in my opinion these "objects" might account for Hidden masses of the universe- and that is a separate quest scientists are embarking on since the last several decades- For those interested to know more about the subject, I advice to read the book: The dark Matter- by Wallace and Karen Tucker). These objects might be traveling at folds and folds of the speed "c"! That must not be astonishing here, since we already admit that what we see in our sky is the past, stars are not really there, we are actually seeing how it used to twinkle millions and billions of years before- so we can stretch our imagination a little more to think of objects felt and seen after some time!!

Maybe that's why the song goes: "twinkle twinkle little star, how I wonder what you are" rather than asking how it was; since of course we are seeing it twinkling in the past. I have to highlight the "proposed" possibility of the photons and Neutrinos being the most outstanding dark matter candidates, simply these photons and neutrinos can account for the dark matter scientists are having extreme difficulty finding in the observable universe. Whereas photons and neutrinos are considered mass less (a photon's mass is considered to be zero, but the energy- E- of a photon is defined by the product of its frequency multiplied by the plank constant h), their mass- if taken into consideration will account for the mysterious unidentifiable dark matter. We might ask why the scientists consider these particles massless? Scientists simply answer: Because it is mass less when it is stationary, very weird indeed. At the end the photon is a particle, but why is it assumed mass less is really very strange. By definition the photon is a particle of electromagnetic radiation that has zero rest mass, zero charge, and travels at the speed of light.

K. PRECLUDES AND EVIDENCE OF SUPERLUMINAL VELOCITIES

K1- The "c+v" measurements or the proofs of superluminal speeds- obtained by radar measurements.

Politics had always played and will always be playing a big role in hiding and closing on these extremely important experimental findings. Unfortunately, this is big fraud and breach of Science and a big cheat to our beliefs. Mostly, these extremely important results were obtained in the former Soviet Union, and were intentionally hidden away from the public. These observations assuring the dependency of light speed on the speed of its source will blow relativity into fine dust - (it will be an alternative solution to the "lorentz transformation" solution of the famous train and embankment experiment of relativity). Considering that light "feels" the speed of its source- the lamp fixed on the train floor, midway between two observers riding on Einstein's train moving forward on a track will eventually reach both observers in the train simultaneously, hence eliminating Einstein's alleged paradox in his train and observers experiment. The light will reach both observers simultaneously in that case, and Einstein's excuse for inventing Relativity will not be there in the first place to start with. Keep in mind that Relativity doesn't hold either way, be it light is affected by the speed of its source or not, this book explains that in detail in several occasions. Hence, the c+v measurements, once confirmed and acknowledged will eventually lead to revision of all our understanding of the nature and behavior of light and other electromagnetic waves and possibly to re-write most of the physical laws as we know it today! But since much of these results are still kept outside the arena, we will turn our eyes away from for a while. However, it is important that we are aware of the existence of these results.

K2- Quasar speeds:

There are measurements of light speeds of quasar components (mostly very strong radio sources, with high red shifts in their observed spectra) that are multiples of the speed of light. Of course many claim that these measurements are illusory, I guess that we already know the reason why .... Because the special theory of relativity does not allow that!!! We already know from "Appendix 1" that Quasar OQ-172 is traveling away from us at around ninety percent of the luminal speed "c", and taking into account the current approximation of the rate of expansion of the universe size; where it is believed that the universe after two billion years will be 15% larger in size than it is today; we can conclude that certain galaxies and clusters of galaxies will eventually reach superluminal speeds. Of course that does not mean that there is no galaxies moving at super-luminal speeds right now, nobody can claim that as I explained in the section about dark matter candidates.

K3- Experiments dealing with Cesium filled chambers:

These experiments had shown light traveling at speeds far beyond the alleged light speed barrier!!! Again Relativity Proponents sacrifice everything including defying logic and blowing common sense in order to cling to their beliefs and use very twisted reasoning – or shall I call it "no reasoning at all" to fight for their wrong beliefs; trying in all means to minimize the importance of these experiments.

K4- U.F.O's

I can just "propose" that; in the cases of the documented reports from all around the world, telling about strange things happening- cases in which people claim to have seen and felt strange things that had no apparent cause, the bodies moving at speeds higher than that of light might be the explanation.... In some cases, people felt the actions of bodies that they have only seen some hours later (when their light reach the scenery of the incident, in the cases that a figure- or a "ghost" is seen without the entity itself) days after, or even lifetimes later..... Bodies "may" be advancing in front of their light figures.... I will not push my speculations more here, as this topic might serve me well as one for a book that might focus on un-explainable phenomenon and U.F.O's.

K5-Stephen Hawking's Interesting Finding

As explained before, and in brief, it is that certain particles are predicted to travel at superluminal speeds for some time to escape black holes' tremendous pull- as Stephen Hawking explains in "A Brief History of Time", and it is mentioned under the headings "black holes' radiations", and described as black holes evaporating.

K6- The Horizon Problem and the Big Bang

The Horizon problem deals with the origin and current size of the universe, this problem can be solved only if we admit that certain parts of the universe had flown apart at rates folds faster than the speed of light. This is referred to as the inflationary stage at the early times just after the Big Bang (Origin of the Universe in a one single tremendous explosion yielding all the galaxies and star systems of the known universe). If any one wants to cling to relativity, he has to abandon the Big Bang picture-theory, which is both widely accepted and proven by the scientific community (but which also had non supporters – who supported the steady state of the universe in which there is Galaxies constantly being created in the intergalactic voids of the universe to keep the universe looking alike whichever way we look it- and that is ironically an experimentally shown feature of our cosmos)). I would like to mention the work of two Russian scientists: Evgeni Lifshitz and Isaac Khalatnikov, who tried to avoid the conclusion of a big bang in 1963. They showed that the universe could have, but not necessarily have had a singularity if General Relativity is correct. Stephen Hawking had also showed that one can avoid the big bang singularity by taking the quantum effects into consideration; and that will bring us to an illogical circle shown at the end of this section. Some more evidence against the Big Bang also came from J. Anthony Tyson and Pat Seitzer of the Bell laboratories, who were using very sensitive charge coupled devices as detectors to catch the very faint light of distant galaxies. Tyson and Seitzer's discoveries showed that some very remote galaxies may be at very early stages of their formation thus contradicting the Big Bang theory for the universal creation. Tyson said "I think that these observations are beginning to constrain the theories. We may in fact come up with a major conflict that will give rise to a completely new paradigm".

However, one of the most powerful proofs for the Big Bang had already been accomplished through the discovery made in 1965 by the German-American physicist Penzias working together with Wilson. Penzias and Wilson were using low noise horn antennas for satellite communications and picked up persistent signals, thereby discovering the cosmic background radiation, which was the single most important evidence for the big bang theory. General Relativity had also predicted the big bang, but it fails to describe the universe at the early moments of the creation, and fails again at describing the proposed Inflation Stage of the universe brought forth by the Big Bang theory. Relativity, in effect, leads to a problem that only quantum mechanics can solve at the moment of the Big

Bang, and fails afterwards when it cannot describe the inflation stage . Supporters of Relativity will not find any way out except by admitting its failure. We indeed live in a world of paradoxes, quantum mechanics holds when relativity fails, relativity thus predicts situations which it fails to describe, and permits its rival – the quantum theory – to take over!!

So in conclusion, it's that either relativity is wrong, or the big bang, or both theories. I think that its relativity which fails here, due to several reasons– which this book is all about.

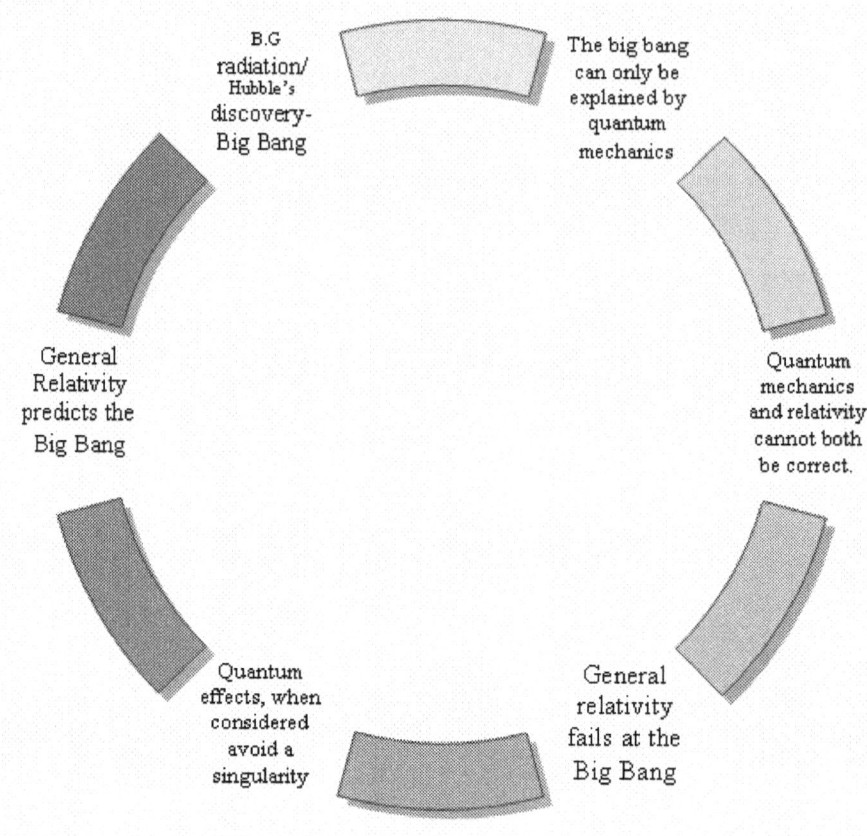

The big bang can only be explained by quantum mechanics

Quantum mechanics and relativity cannot both be correct.

General relativity fails at the Big Bang

Quantum effects, when considered avoid a singularity

General Relativity predicts the Big Bang

B.G radiation/ Hubble's discovery- Big Bang

K7- Photon Pairs

Experiments in the field of Quantum Physics had shown that photons emitted as pairs are eventually coupled by their quantum numbers, and that coupling emerges at velocities exceeding four times the velocity of light and may even be infinite!!! What is more interesting is the ability of these experiments to experimentally show the absolute speed of the laboratory through space with a good accuracy (the same as proving the existence of the universal Aether- that relativity forbids)!

K8 – Tachyons

To be able to amply cover most of the subject of super-luminal speed possibilities, the Tachyons are an ultimate candidate for breaking the top speed limit "c". These hypothetical- till yet un-proven particles are able to break the speed "c" and the Lorentz Transformations did not exclude their existence.

K9- Gravity

When we already know about the COBE results of imaging the point of creation- the big bang, the alledged Reynmann space of Einstein proves an illusion, and we are back to our absolute space idea of Sir Isaac Newton-where Euclidean coordinates return to work. We are also back to the idea of gravity acting as a force, and not acting through the warped space time as Einstein used to described it; hence, we are back to instantaneous action at a distance, and of course the infinite speed of the action of gravity between celestial bodies of our cosmos. That is actually a "double fall" for relativity, in both its Special and its General versions! The speed limit "c" is broken and the "warped" space returns to its initial Euclidean structure- in much the same way our ancestors saw the universe and knew it is all right! General relativity predicts that Gravity is manifested through Gravity waves. Anticipated by Einstein himself, these gravity waves have not been proved to exist experimentally till today. Gravity waves will have to travel at the same speed as light. Gravity waves may consist of gravitons- which had also not shown to be existing. According to General Relativity, these gravitons are said to be the means that gravity applies itself.

5. ON RELATIVITY AND SIMULTANEITY

...Figuring out the problem is sometimes more important than finding the solution-which might be very trivial after the problem is known... But many times finding out the problem is a big problem in itself....

Einstein wanted all observers to see simultaneous events taking place simultaneously (his train and embankment model experiment- in his book "Relativity"), and that's why he uses the famous Lorentz transformations together with his basic wrong assumption that different observers must all measure the same speed of light to show that time must be dilating and distances must be shrinking!!!

Yes, he has indeed created this jargon because he couldn't realize that observers need not see perfectly. After all, what is wrong in observers seeing different things differently? That is nature: humans are not perfect, they don't have eyes and ears everywhere-their senses are restricted. Albert Einstein unjustifiably shows that time dilates using Lorentz transformations, concluding that time slows down for fast moving objects, and lengths shrink in the direction of motion- with respect to the moving observer. I quote Einstein's words from his book –Relativity, we have to notice his basic wrong unjustifiable assumption that Light speed "should" be measured the same by any observer: "Can we conceive of a relation between place an time of the individual events relative to both reference bodies, such that every ray of light possesses the velocity of transmission c relative to the embankment and relative to the train? This question leads to a quite definitive positive answer, and to a perfectly definite transformation law for the space-time magnitudes of an event when changing over from one body of reference to another." Stated otherwise, and by referring to Einstein's train of thoughts and reasoning, and after reading his book "Relativity", we find out that Einstein had a skewed understanding of "simultaneous events as seen by observers". It is the product of his unjustifiable "belief" that all observers must be measuring the same velocity "c", whereas all he had as evidence of the invariance of luminal speed is his awkward explanation of the null result of the MMX!

Most of the readers might be wondering...Is it possible that was Einstein misled? I like to reply, yes he was, and the reason is that he was trying to find a very difficult solution to a very simple problem, he created confusion and paradoxes as a solution to his feeling of un-easiness when he did not understand what is "Simultaneous" .This was triggered by his mis-conception of observers, I can easily imagine that he did not feel the real sense of the word "observer" and what an observer should be "observing". An observer is not un-mistakable; nature doesn't have that much of a respect towards any particular observer, and might be deceiving him. We should see no problem in different observers measuring different physical quantities differently. Unfortunately Einstein had discomfort in that, I personally feel very relaxed when I think of this problem that way. In that regards, I have a powerful argument I will present.

Imagine an observer-A- living on a planet in a galaxy several million light years away from us (it would take light several million years to reach Planet Earth starting from there), and another observer -B- living on earth. Imagine that when you are reading this sentence, observer A witnesses a comet falling on his planet right now, any body will know that observer B, on earth- (who can be you- the reader) will know about that incident after light reaches him, actually it will be your great great great.......... grandchildren, who might not be looking like you anyway, nor might they still be on earth – if it remains to exist then. Yes it will be your great great grandchildren who will witness the comet falling on that planet.

Very clearly, observer A and observer B witness the same incident at different times, very clear to us, but just slipped past Einstein's conception. Einstein was very clever indeed, but missed on that. That does not contradict common sense nor does it incur problems in our perception of simultaneity of events, I only wish it did the same with Einstein back then. Einstein, till yet had lead mankind to a hundred years of confusion, he invented two vast tell-tale theories to cover up for his mis-understanding of Relativity and Simultaneity. For this, in my opinion, he had delayed scientific progress and is still doing so, at the same time he had provided material for science fiction stories and films. Albert Einstein had fueled science fiction film makers' imaginations to create science fiction scenarios. Moreover, Relativity has lead many nations to embark on building mega scale projects built on the mentality of providing support to relativity and prove its predictions whereas relativity must be brought to true and objective testing..

Very clearly, Einstein had used serious assumptions; which, coupled with his misunderstanding of Simultaneity lead him to invent Relativity as we know it today.

6.   ATOMIC CLOCKS AND THE IMPERFECT TIME KEEPING - PHOTONS DECIEVE EINSTEIN

What philosophers and other great people had said about time:

To see a world in a grain of sand,
And heaven in a wild flower,
Hold infinity in the palm of your hand,
And eternity in an hour- William Blake

Time will reveal everything. It is a babbler, and speaks even when not asked- Euripides

Tempus fugit "Time Flies"- Ovid

O, call back yesterday, bid time return- William Shakespeare

As if you could kill time without injuring eternity- Henry David Thoreau

What Madonna said about time:

" and Time goes by so slowly..."

The last word, however, should be kept for science:

A. BLACK HOLES AIN'T SO BLACK- OR RELATIVITY AIN'T CORRECT?!!? "EINSTEIN MAY NOT BE ABLE TO SLOW DOWN TIME ANYMORE! "

Returning to the particles escaping the black hole (Stephen Hawking), and if we believed that time slows at higher speeds- assuming Relativity is correct, then these particles actually go backwards in time since that is super luminal speed (which by itself contradicts relativity). Black holes again constrain Relativity, Relativity fails twice in the vicinity of these ultimate space carnivores, which can gobble up whole stars at once. Relativity does not allow superluminal speeds, and Relativity will imply backwards motion in time if we close an eye about the superluminal speed particles. We are faced with two alternatives, if we are confident that time doesn't go back: either to admit relativity wrong (time doesn't slow down for fast moving particles) or to disagree with Stephen Hawking's theory (that black holes are supposed to radiate- tossing off particles at super luminal speeds- which is widely accepted)! I think that our logical choice should be and will be to abandon the implications of special relativity!!! Stephen Hawking has actually proved that Black Holes must be evaporating, but Einstein had just brought forth the special relativity that we find very little truly supportive evidence to, but in which we find many paradoxes and many contradictions not to mention the affront it brings to common sense and logic.

B. IT'S STILL A NEWTONIAN UNIVERSE WE LIVE IN

We have every right to conclude that time does not slow down. We still live in a Newtonian Universe, the Universe hasn't changed to an Einstein's Universe where someone can go back in time to kill his grandfather then return to live his normal life, or in which one of two twins can decide to decrease the rate of his ageing compared to his brother's by going on a high speed journey for a couple of years. Einstein's universe is a place where things contract while moving at speeds near the speed of light, and where mechanical clocks should go slower, while there is no single reason that proves so (all what Einstein's supporters have is the theory that atomic clocks "might" be going slower at high speeds.. More than that it is the universe in which all heavenly bodies were scrammed into a small tiny singularity a time of the big bang(which I personally face difficulties to imagine), and at the same time a small grain of sand might have the weight of more than all our solar system   if *accelerated towards the speed of light...........* Alleged "proofs" that Relativists typically use as evidence for time dilation like the longer life of particles- shown in the chapter "Where is the truth" prove to be mere biased explanation of observation and not real evidence of time dilation, time dilation is just a "proposed" explanation when scientists do not actually know the real reason holding particles from disintegration at high speeds.

C.ATOMIC CLOCKS ARE SAID TO BE SLOWED DOWN BY HIGH SPEED AND BY GRAVITY

Supporters of dilating time theory move forward to use the proposed imperfect time measurement of a not so perfect time keeping machine- the atomic clock to prove that time indeed does slow down..... Again, this is wrong interpretation of experiments. Atomic clocks which record the passage of time in the most reliable manner slow down when moving at high speeds comparative to the speed of light and also slow down under influence of higher gravity; therefore time is said to slow down. Let us see if that is common sense at all: As per the supporters / proponents of the time dilation theory, their argument is that photons, which bounce back and forth between two opposite plates in an atomic clock, which record the number of bounces and hence, the time passing will have a longer path to go while clocks are moving. When an atomic clock moves, in a certain direction, the reciprocating movement path of the photons will turn into a longer path resembling a parabolic wave, and baring in mind that photons travel at speed of light, it will take them longer to hit the plates, and hence, our moving atomic clocks will start showing slower time passage....! Under stronger gravity, we know already that photons prove to slow down, as in the case in the vicinity of black holes and under strong gravity, so it will take them longer to travel between plates, and also, the atomic clock runs slower..... A high school student will jump up and say Atomic clocks may be wrong devices for measuring absolute time, and you guess – he may be ultimately right, and the experts are wrong!!

Stephen Hawking in "A Brief History of Time" mentions the water tower experiment carried out in 1962 confirming that the atomic clock fixed at the base of the tower ran slower than the other fixed on top of the tower, and he agrees that this is in strict conformity with special relativity!!! The reasoning is that stronger earth gravity at the water tower base is responsible for slowing the atomic clock positioned there with respect to the one placed at the top of the tower... While this slow down actually happens as shown by the experimenters, I will restate that the error is in the misinterpretation of the results in favor of relativity(with all our admire and respect to Stephen Hawking – even if that is not the main subject of his book, but he shows that he also is mislead here) . If an atomic clock runs slower, it means that the atomic clock runs slower and *is therefore not reliable, and not that time slows down!*

Before humanity started using atomic clocks, traditional mechanical clocks were used, an elementary level student will guess out that a mechanical clock does not use photon movement for time keeping, so common sense will say that a mechanical clock will not slow down - as there is no reason for that (especially at constant speed motion, or no-force motion), we have the right to think that humans have not till yet, invented the "ultimate time-keeping device", the device that will not slow down or otherwise while moving at comparatively high speeds or under the pronounced effects of gravity. Humans don't, but God does count time properly......Somewhere out there!

We may infer that time does not dilate....Not to mention that man, supposedly, still has all the time to think, and to invent the perfect clock machine - the clock that doesn't depend on photon movement to record passing time!

D. EXPERIMENTS INVOLVING TIME AND HOW IS IT INTERPRETED

The Hafele-Keating Experiment

It is said that in 1971, Joseph Hafele and Richard Keating flew Cesium Beam clocks around the world in a commercial Pan Am jet. After several long hours of flight, they compared the Jet-mounted clocks readings to the Earth Bound Clocks, and they claim they found very miniscule discrepancies – a few hundred billionths of a second, in close agreement with Relativistic calculations.

This experiment was intended to show that atomic clock movement was affected according to Einstein's Relativity principles and was the experiment relativity proponents use as their strongest proof of time dilation. It is "said" that the Cesium beam clocks behaved as the relativity theory assumed. What's on the other side, we must hear the story: Engineer A.G.Kelly obtained the value numbers of that test conducted in 1971 and showed that the original test results did not show at all what was mentioned in the paper of 1972!!! (Any one interested can refer to A.G Kelly's paper for further reading on that subject)- If this miniscule time discrepancy is the strongest evidence for the relativistic effects, Relativists are a long way from proving

Relativistic effects on time beyond doubt!

Mu-mesons

Richard Feynman, a leading 20th century theoretical physicist and teacher explains how time dilates, and how he thinks time slows down for particles like mu-mesons. Usually mu-mesons are particles that disintegrate quickly with a life time of 2.2x10-6 seconds, not sufficient for them to reach from the top of the Earth's atmosphere where they are created to Laboratories on Earth. Yet these mu-mesons are found to be reaching down to the surface of the Earth!

Feynman explains that we don't know what causes the mu-mesons to disintegrate and behave the way it behaves, but relativity predicts that these particles "should" behave the way they should, so that is ample prove for time dilation-or time slow down- for these particles! So in effect Feynman is admitting that scientists actually DO NOT know why the mu-mesons last longer, and therefore conclude that they last longer because time slows down relative to these mu-mesons!

That is really a very clever way to explain observations when we do not know the real cause for what is happening, and that clearly shows Feynman's support for Special Relativity. So the new motto in science is not to search for the real reason behind observations, but rather to try our best to fit whatever we have to suit theories that must otherwise be put to true testing. I have to mention here that the possibility of time actually slowing down "might" be one of many reasons behind mu-meson apparently longer lives. Why do Scientists like to stick to "this" explanation? Simply because it suits Special Relativity. In that case, are we really searching for the true causes? Probably no. But why? Because some don't like to know - since they are not interested to start with!

E. WHAT TIME ARE WE IN?

What a time we are in? It is a time where we can make time pass slowly in some places while we keep it running fast in other places.... It reminds me of frying potatoes, some fry well in the pan and some get burnt while some stay raw, and then can we call it a perfect dish?!?! What we should conclude is that time doesn't slow down or change pace; it is neither affected by high speed nor by gravity even if our most accurate clocks are affected. On the contrary, even though light posses this extreme speed, when used in moving atomic clocks to record time, it proves that it is a wrong tool when motion or gravity is involved. The perfect time keeping machine here will be a super and eternal power sitting somewhere stamping its feet on the matter beneath on a great heavenly body we have not found till yet; at an ever unchanging pace!. That eternal power does not depend on photons bouncing between the plates of a manufactured atomic clock to measure passing time.

7. PARADOXES OF SPECIAL RELATIVITY AND THE FAILURE OF GENERAL RELATIVITY

...All humans are born smart, but the lazy and un-interested are the smartest in getting excuses ...Tarek S. Ahmadieh

A direct and simply recognized paradox of special relativity is the fact that one of its direct consequences is that in a body of two systems, if the first body was moving against the second, then the second body must be moving relative to the first also. Special relativity therefore predicts that body A shrinks with respect to body B (as seen from A) and body A shrinks with respect to body B (as seen from B), and so we have bodies shrinking not because it "should" be shrinking – due to laws of physics, but rather because it is being looked at- or "watched" by other bodies (observers) in other systems –ridiculous indeed!!! So if two astronauts are moving in opposite directions to one another in empty space, at speeds appreciable to the luminal speed c, each would see the other shrinking in size, and each would feel that the other is acting slowly since the other's time is slowed down while of course the mass of the moving astronaut (which are both) will grow- or both astronauts will swell (without noticing for sure). But wait a minute, Dr. Einstein assures us that each of the astronauts will feel nothing strange taking place in his world – from his point of view because each astronaut has his right to claim himself stationary and that all other bodies in the universe moving relative to him!!!! This is indeed a crazy outcome of special relativity.

General relativity, and while claiming the explanation of the bending of light due to the sun's gravitational field, fails to explain gravity in most of the cases especially in zones of strong gravitational grips. However, it has been shown previously that the light bends due to refraction in the sun's atmosphere, and not due to gravity, thereby falsifying the claim of general relativity. I have to cast light on the fact that Einstein was always un-easy about the ability of His General Relativity to hold true in the strong gravitational fields- whereas the followers are the strong believers and the ever obeying slaves of relativity. General Relativity has not till yet proven universally applicable(it fails at black hole event horizons, it fails at the moment of creation, it fails at strong gravity fields). I recall that for a certain theory to be proven correct, it must be always up to the challenge, it must work everywhere, the more constraints it has, the less reliable it becomes, and the less the faith we can put in it! We should voice out a clear message to relativists: General relativity fails to explain the true source of gravity as well as the means gravitational forces are applied through. It has been also shown that the space we exist in is a Euclidean Space and not a warped Continuum of space and time interacting with each other. It has been shown that the perihelion advance of Planet

Mercury is explained by Paul Gerber in 1898 before General Relativity and even before Special Relativity!

Well, I think that General Relativity will remain surviving in the heads of those who usually look for complex jargons and beliefs for their in-ability to know the real cause behind things- the knowledge of " how much" and not the knowledge of "why things are the way they are".

Science is a description of the truth- and who have never asked "why" or who are not very interested in knowing "why" are just followers – forever trailing behind, while they seek their benefits and interests, and not the truth. Science is not dogmatic beliefs and it is not personal interests and tastes.

It is believed that Einstein used the complex Reymann Geometry just because it provided him with the "extra variables" he needed to mask the face of his General Relativity, the equation- "tensors" of which, had failed to be described by the simple Euclidian Geometry. Indeed Einstein created complexity and introduced superficial dimensions just when his General Relativity could not be described by the simple Euclidean Geometry! He must have been right in being un-easy of the validity of his General Relativity, for he is the Cook and he is the only one who knew about the secret recipe and the suspicious un-healthy ingredients he had used, is this the tasty dish of truth we seek? No, sure not.

At the end, we recall the three "classical proofs" of the

General Relativity which are:

1   The gravitational Red shift- which is really a proof that Light is waves and thus needs a medium- which means the existence of a medium - or the failure of special relativity to start with.

2   The deflection of light- already shown to be due to refraction and not due to gravity of the sun.

3   The perihelion advance of Mercury – already described before and shown to be explained by Paul Gerber before Einstein knew how to read and write.

8. A PROBLEM OF ETHICS

"Science is a wonderful thing if one does not have to earn his living at it."- Albert Einstein.

It is believed to be said by Albert Einstein himself, It surely is correct, for when somebody is working something he is not paid for , he will be doing it from his own desire, he will probably excel at it; whereas when he does something for a living, he is probably working to justify his salary. An employee need not necessarily believe in what he is doing, he might be working because that is what he "has" to do for a living!

It is world wide known that significant numbers of talented students who are very clever at mathematics and who have really high Intelligence coefficients fail to master the subject of relativity, I suppose that we know the reason why now, it is sure that the problem is not in the students, it is in the subject itself and in the brain washing they are supposed to subject themselves to. Their in-ability to accept relativity represents their intelligence and is a measure of their common sense. As for me, I strongly believe that most of the interested people who say they digest Relativity are either getting it wrong or must have a really skewed sense of common sense or no common sense at all, in the case they did not find anything strange and annoying. It has become a standard practice at the high university levels and at research institutions that any works that can threaten relativity are being discarded and put away, the highest authorities at the Graduate Levels and at the PhD levels have become committed to cut the way on any work that may be considered offensive to Relativity. Relativity is a sacred temple after all. These people have become enemies to scientific progress and an obstacle in the search for truth!

I think what we basically have right now is a rapidly growing segment of the scientific community that is accepting the "fact" that relativity is witnessing the end of its glory. Others are very skeptical about relativity and even though sometimes very convinced of its falsehood, but they never admit that openly because by doing so, they might loose their life long careers and jobs. A not so small part of the concerned scientific community have a very dogmatic attitude, and know that by admitting Relativity untrue, they might face serious ridicule, and therefore feel very embarrassed for their personal favors and benefits and are careless for the advancement of science. Supporters of Relativity, on the other hand, are either doing that to stay in their positions, or because they feel that they had lived with, and accompanied this "vice- of believing in Relativity" sufficiently a long time so that their minds are sufficiently twisted and tuned to be able to see as white what is black and as black what is white!!!

We have very powerful political reasons keeping relativity in its place which is always present to suppress any possible attempt to judge relativity, yes indeed there is governments behind that, there are interests, most importantly there is money which the most are running after. Many have already un-masked relativity. Late Peter Beckmann of the University of Colorado published his theory refuting relativity and established a newsletter called Galilean Electrodynamics. Herbert Eves, an engineer at Bell Laboratories, conducted his personal experiments in 1930's to prove the existence of Ether, he summoned by accusing Einstein of being a great paradox swallower defying Logic and common sense. Stephan Marinov had threatened to immolate himself in front of the British Embassy in Vienna because the famous science magazine "Nature" refused to publish his proofs against relativity. Ruggero Santilli, an Italian physicist, had written a book "Ethical Probe on Einstein's followers in the USA- he speaks about conspiracies against his attempts to conduct research on his theories to disprove relativity at Harvard".

Each day, more and more scientists and researchers are discovering the true reality of relativity and what fraud it is to science and common sense. Relativity, unmasked, reveals a theory based on nothing but paradoxes, it is reached upon by a series of invalid assumptions, piled one on top of the other to reach surprisingly erroneous conclusions which any scientific mind had trouble ingesting ever since its birth in 1905.

H.E Retic, who purposely conceals his full name, tells his experience when explaining his proofs against relativity to a specialist in that field- the field of relativity. This "specialist" not only did accuse Mr. Retic of being an ignorant, but went on rage and described him as "a dangerous heretic who must be suppressed..." Mr. Retic clearly and un-doubtfully understood that this "specialist's" quasi religious beliefs were threatened. What a shame indeed, when specialists in relativity become deadly guardians of the temple of relativity, shooting away at anyone who passes close! They are surely not abiding to their masters words anymore, for wasn't it Einstein who said that the search of truth must take precedence over the teachings of the established authorities regardless of the prestige of that authority?

Science is our experience with nature, if we transform a particular field of Science to a Religion, that would be the dooms day of science, Science is investigation, experimentation, and endless search for truth. On the other hand, religion is faith and belief in what we assume as God's commandments, religion thus provides us with the truth in a different relative perspective-depending on the individual's beliefs. A true scientist should not be loyal to an idea; his loyalty should be to the truth. Through the ages, science could not have advanced if primitive beliefs regarding the structure of the universe were kept outside the circle of objective scrutiny.

On the other hand, I think that relativity will not meet its logical fate- the history of dead theories that fast.... And the reasons for that are many; Many institutions' own existence might beat the purpose for which it was brought to be; that is, if Relativity was admitted to be false. Many world authorities still regard Relativity as a futile valley for wasting people's income taxes on tests and researches relating to relativity and its proposed outcomes.

9 "PROBABILITY OF EXTRATERRESTIAL LIFE" and "OTHER UNIVERSES"

...To bring world peace, let us focus on the things we have in common, while appreciating our differences... We better be prepared to encounter the Aliens-who will sure be very different from us... they might also be different from one other... but for sure, they will never be interested in solving our problems- Tarek S. Ahmadieh

Once, astronomers believed that a certain patch of the sky is empty; using their available equipment they did not find anything, so they assumed that the patch of the sky is intergalactic voids (voids between galaxies). In 1983, astronomer J Anthony Tyson and Seitzer working at Bell laboratories in New Jersey began a survey of twelve big patches of the sky assumed before almost entirely empty. Cutting their story short; they had found, wit the aid of their four meter telescope and taking advantage of their charge coupled device employing a grid of thousands of light sensitive cells that are able to count individual photons. It is said that they counted around 25,000 faint light sources. What was believed to be entirely empty regions of space housed around 25,000 galaxies!!!

We live in planet earth, part of the solar system that is also part of the larger structure called the Milky Way galaxy, and the patches observed by Tyson and Seitzer showed evidence of an enormous number of galaxies like our own Milky Way galaxy. I would like to restate that we merely saw and proved these galaxies' existence; while each galaxy appeared as a small dot of light on the sensitive grid, it actually houses millions and billions of stars, many like our sun, many smaller, and many larger. But at such a large distance all appear as a tiny patch on our detectors. What do we know about the structures of these galaxies, we know very little indeed. Till yet, all the human endeavour, including our probes to one of the nearest planet –Mars, had not been conclusive regarding existence of life. We are still a long way to be sure our solar system doesn't house extraterrestrials and of course very far away from concluding the same regarding our Milky Way Galaxy. We are still now discovering new plants and insects on our home planet; we are by no doubt, a long way from knowing what other galaxies will probably be holding as surprises for us. We know very little about the closest large galaxy to us, like the Andromeda and the Triangulum Galaxies, let go the very far away galaxies at the edges of our universe, more than 10 billions light years away, and which appeared as a mere dot in our most light sensitive equipment as explained before. We ultimately know nothing about possibilities of life in these galaxies.

Galaxies may contain thousands to trillions of stars, many stars, on the other hand may be orbited by planets which might house life in any form. One of the first planets to be discovered is 51 Pegas; found in 1995 orbiting, the reason that planets are hard to find is because their light is diminishingly faint when compared to the strong light of the stars they are orbiting. As we probe deeper and closer into the far edges of the universe; millions and billions of planets will eventually be discovered enhancing the probability of life existence. Let us consider the chances of finding life at other places in universe assuming that life can only develop on planets like the planets of our Solar System.

Let us begin by assuming that the probability of life existing on a certain planet is α (α being a decimal number less between 0 and 1, we will consider it to be equal to 1/1,000,000 or 0.0001% (one over ten thousand of a percent); now the probability (β) of life not existing on a certain planet will be such that:

$\beta = (1-\alpha) = 999,999/1,000,000$

As the number of discovered planets grow, and if we consider that the probability of life existing on a single planet is independent of probability of life existing on any other second planet, so the probability of life not existing on (n) planets (β') = (β) raised to power n,
And as n $\rightarrow$ ∞ (infinity)
β' will tend to zero,
I.e. the Probability of non-existence of life elsewhere in the

universe will progressively diminish as we are sure there are billions (and maybe trillions) of planets yet to be discovered.

I would like to mention the project SETI (search of extra terrestrial intelligence) around the globe using radio telescopes, which pick up and broadcast signals. SETI aims at finding evidence for extra terrestrial life. Scientists have sent massages (such as the Arcebo message explaining life on earth) to the other worlds, and thereby await extra terrestrial reply- of course if we do succeed in making contact with aliens, it will be the single biggest success for mankind.

Neither the Big Bang theory nor the steady state theory excludes the possibility of other universes' existing simultaneously with our known observable universe. I personally had made myself acquainted with a multi universe- or the "multiverse" idea and find the picture very nice, as there is no theory that excludes or disproves this proposition. So is it one single big bang at the start of time? Maybe none, maybe many, and maybe we have one big multiverse, some universes of which are contracting, some being steady, some expanding, while some others coming to birth while you read this sentence! Who knows? Probabilities are never ending here. Speaking of the Big Bang theory, which I personally feel very un-comfortable with, there exists the so called Horizon problem, which brings strong evidence against the Big Bang Theory if it could not be explained ... The problem is that how could it be that cosmic heat radiation has reached from the horizons of the universe to many other places of the cosmos – knowing that the Universe spans around 28 billion light years across (a light year is the distance light travels in one year) .It might be that radiation is going faster than the speed of light or might be that the universe has not started with the Big Bang. At any case, one of the theories might be in trouble- the Big Bang or the Special Relativity. I think it will be the Special Relativity-already too weak to withstand outgrowing evidence against itself.

10. WHY PI?

"... and if It can be done better, then we must find the best way to do it..."-Thomas Elva Edison

Why does nature choose numbers such as "Pi"? What does it mean when the two entities (radius and circumference or radius and area of a circle) are related through an irrational number such Pi? In the same time, there are simpler relations in physics, such as the one relating the force (F) accelerating a body of mass (M) by acceleration (A), that is: F= MA, such a formula is simple, and has no irrational numbers- no number at all ... It is a beautiful formula indeed....

The question that poses itself is "why should there be a constant in formulas such as E =mc2 or A=Pi R2" in the first place... There after, one can ask why did nature choose "C" in the first equation and why did it choose "Pi" in the second? Figuring out the answers would be a very challenging task, if ever it had to be figured out. I can, however state that the equation relating the radius to the area of a circle will forever feature Pi, as Pi is proven to be the ratio of the circumference to twice the radius, and the ratio of the area to the square of the radius. This proves out to be a reality, something we did and will not play a role in .... Nature had chosen Pi without consulting us.

On the other side, looking at this tantalizer in a special

way – the child's way; we may speculate that if Pi was found to be necessary when we have to relate the area to the radius or the circumference to the radius of a circle, wouldn't there be another way round in which we relate these properties of a circle without having irrational number constants such as Pi?? Trying to answer the above question, as silly as it may seem; had taken me more than 15 years of continuous speculation, At the end, I am fortunate enough to be able to write down the following sentences:

The area of a circle is = half the circumference multiplied by the Radius, $A = CxR/2$ .... Very clean and beautiful.... with no irrational numbers!!!!!!!!!!! Shall we stop using the famous $A=Pi$ multiplied by the square of R? and re-write the mathematical relation as: $A=CxR/2$? Of course not; for mathematicians, engineers, surveyors, students, draftsmen, Carpenters, Astronomers, and teachers will all use it when they are given the Radius in numerous studies and applications. As long as we don't have the measure of the circumference, we cannot calculate the area based on the formula $A= CxR/2$. Technology hasn't provided, till yet, a flexible specially designed Curved Scale, which we can use to physically measure a circle's circumference- with high accuracy, enabling us to derive the area by multiplying the circumference C by $R/2$; without the need of approximating the area through the use of approximations of the irrational Pi...Rather, we can very simply place a string over the circumference and then straighten out the string and measure its length. You may be asking yourself the essence of the talk above, and why do we have to bother after all, but the inference drawn here is rewarding. I have presented two versions of the mathematical relation between 2 element-features of a circle;

   1- The known $A = Pi \times R2$, including an irrational "Pi"

   2- The "inferred" $A = C \times R / 2$, which "does not" relate the element-features of a circle through an irrational number.

The Deduction

While the two formulas shown above both do relate elements of a circle through a mathematical relation, the second proves more "natural", meaning that the area A of a circle is more "naturally" and "mutually" related to C and R than it is to the square of R since the latter has to include the irrational number "Pi".

The world is heading towards simplicity, towards describing all observable processes through simple clear theories such as the string theories, which show that all processes of the world originate from action of vibrating strings which are indeed the most elementary and smallest building block of everything in the known world, while the oscillating strings are considered the "basic" constituents of all matter around us. Our aim therefore should be to "discover" the "purest", the "easiest" or the more "natural" formulas relating physical quantities and features, but while these natural formulas may be very easily discoverable, it may also be very hard to deduce. World wide applicable formulas such as the formula measuring the force of gravitational or "weak" attraction between any two heavenly bodies, and which include constants such as G, the universal gravitational constant, which equals $6.672 \times 10(-11)$ N m2/Kg2, are not formulas at the "purest" state yet !!!!

Nature does dictate simplicity, it does not want irrational numbers, and does want us to reach ultimate truth of discovering ultimate pure formulas. *Most important of all, nature wants us to "identify" which physical entities to relate in a formula, which are better related to each other... A good starting point would be the currently available and adopted mathematical relationships, which we should reform and refine constantly to reach ultimate state of relating "naturally" related physical entities.*

Natural mathematical relations would therefore possess a natural reality, manifesting it self in naturally related entities, rather than awkwardly related entities that would impose constants such as Pi and G, among much others. I see no rational reason standing behind nature so that it dictates numbers such as G, Pi, K (the electric constant). On the other hand it was our attempt to relate "awkwardly" relatable entities that lead to mathematically occurring constants in relationships written on paper. In our new endeavor for finding a universal "natural" set of mathematical relations describing all physical phenomena, we might find new physical entities- that may *represent features/ entities of the world that may or may not be related to known physical entities. We might, on the other hand, discover new units of measurements, and these units may be indicative of new basic blocks underlying the eternal marriage of matter and its constituents, forces in all its forms, and time with its implications.*

The newly proposed theory of everything (T.O.E), is described as being the ultimate, final, and coherent explanation of all processes. It would need no further simplifications, since it would be based on the newly proposed string action properties. While these strings are shown to be the basic indivisible basic building blocks of everything in our universe, I quote from Brian Greene's "The Elegant Universe":

Every Particle of matter and every transmitter of force is a string whose pattern of vibration is its "finger print". Because every physical event, process, or occurrence in the universe is, at its most elementary level, describable in terms of forces acting between these elementary material constituents, the string theory promises a single, all inclusive, and unified description of the physical universe: a theory of everything (T.O.E).

While the world is going towards such a theory; clearly, strings are being considered rather than particles such as electrons, protons, quarks, neutrons, photons, or even waves and wave features and entities. *We had therefore started a journey on the track of finding the new "natural" and ultimate* mathematical relations relating basic "naturally" relatable entities that will eventually lead to a cleaner, state of the art- relations relating what should be related in terms of physical entities in the proper manner it should be related in... In doing so, our world will eventually become free of "irrational calculated constants, and unwanted complications resulting from relating un-relatable entities in non-simple context.

11. MY AIM

... Listen to the experts, let them explain why it cannot be done, then go and do it...

The most recent battle is between quantum mechanics and Relativity, quantum mechanics has widely succeeded in describing the sub-atomic world, where Relativity failed to do so. Quantum mechanics proved to be able to give insight to the first moments of creation- if there was one- but relativity couldn't. Lately there is evidences that the universe, on its widest scales has an intricate structure, Galaxies and Clusters of Galaxies are not just haphazardly distributed in space, but rather exist in formations which reinforces the idea that quantum mechanics might be able, as well, to describe the universe at its largest scales, in the same way it described matter at its smallest scales.

All I wanted to show is that relativity; in it self, is self contradicting. Conclusions of relativity are, by them selves contradicting of the very same basics it is inferred from.

To this, we cannot be always reluctant, and to reality, we cannot turn our backs. We all pay respects to Einstein, we ought to continue his mission: search for answers of (why and how), and always ask the invaluable "why".

When we lose the ability to ask this "why", we will lose our ability to criticize, we start accepting wrong ideas, and by accepting it, we might commit vice and not ask why, perhaps it is because we are taught to do it, and we never asked why in the start.

We are always learning and adding to our mental library, but rarely are we questioning what we are being taught. Thinking "why" is steps ahead from just "knowing how and how much" or just "learning". Most of us are thinkers, what we think about are our daily concerns and matters, not universal matters that we are involved in, but may not be conscious of, just because we are doing well without bothering at all.

APPENDIX-1

Ohio State University Radio Observatory, Columbus, Ohio

By Jerry R. Ehman

In the case of OH471, the optical spectrum revealed a very large red shift (Doppler shift) due to the expansion of the universe; in fact, it was the first object known to have a red shift greater than 3 (specifically, 3.40) putting its distance at about 90% of the way to the edge of the visible universe (it was called "the blaze marking the edge of the universe"). Many other Ohio Specials also led to interesting discoveries (e.g., OQ208: a galaxy at about 1 billion light years distance; OJ287: an erratic, rapid, violently variable quasar; and OQ172, a quasar with the even higher redshift of 3.53). In 1976, three years after their redshifts had been determined, OH471 and OQ172 were still the two objects with the highest known redshifts (at distances of about 12 billion light years).

APPENDIX-2

Red Shift, symbolized by (z): The amount by which the wave length of light from a receding object is lengthened (i.e. moved to the red) by either the *Doppler Shift or the expansion of the Universe. Red Shift is calculated by the formula $z = \Delta x/x$, where x is the original wavelength (as measured in the laboratory) and Δx the observed change in wavelength. A red shift of 0.1, for example, means that the light has been red shifted by 10% in wavelength, whereas a red shift of one means a change of 100 %(i.e. a doubling in wave length). At red shifts less than about come into play 10% , z is related to the velocity of the object, v , by the simple expression z=v/c, where c is the speed of light. At large fractions of c, the effects of relativity come into play, and the red shift must be calculated from: $z=\sqrt{[(c+v)/(c-v)]}-1$ (Oxford Dictionary of Astronomy)

APPENDIX-3

The velocity of light c through a medium having the properties: Magnetic permeability = μ

Dielectric constant = ϵ; is

$c = \sqrt{(\mu * \epsilon)}$;

Similarly, the velocity (V) of sound waves moving in a solid body having the properties:

Elasticity = e

Density = d; is

$V = 1 / \sqrt{(e * d)}$

Another proposed expression showing the speed of light as a function of the fine structure constant, the planks constant(h), the charge of the electron (e), and the dielectric constant (ϵ) is:

$137 = \epsilon * h * C / e^2$ (H.E.Retic)

APPENDIX-4

GTR Tests - The Pound-Rebka-Snider Experiment

Dr. Adrian Sfarti

1. Abstract

Einstein predicted a change in the energy of photons in the proximity of a gravitational field, the change being directly proportional with the distance from the gravitational source. In the early 60's Pound and Rebka1 have set to verify Einstein's prediction. The experiment was reprised with even higher precision by Pound and Snider2.

2. The Pound-Rebka Experiment

The Pound Rebka Experiment

The experiment was set up in the Harvard tower. The Harvard tower is just 22.6 meters, so the fractional gravitational red shift between the light frequency v at bottom of the tower and the frequency v0 at the top predicted by GRT given by the formula:

$(\upsilon-\upsilon o) / \upsilon o = gh/c^2$

is 2.45 x 10-15 . In (1) g is the gravitational constant, h is the tower height and c is the

speed of light in void. Pound and Rebka used the Mossbauer effect with the 14.4 keV

gamma ray from the iron-57 isotope that has a high enough resolution to

detect such a
small difference.
In reality Pound and Rebka measured the energy difference $\Delta E = \eta(v - v0)$ (2),
η being the Plank constant.
From (1) and (2) we obtain:

$$\Delta E/E^2 = (v - vo)/\ vo = gh/c^2$$

Comparing the energy shifts on the upward and downward paths gives a predicted
Difference:
$\Delta E/E$ down - $\Delta E/E$ up = $2\ gh/c^2$ = 4.9 x 10^a, a = (-15)

The cleverness of the experiment lies in the fact that it sidesteps the
measurement of the frequencies at the top and of the bottom of the tower
and it replaces such measurement with a very high precision energy
measurement and here is where the Mossbauer effect comes into play. The
measured difference was 5.1x10-15. The results were improved in the
subsequent experiment co-authored with Snider2. The tests were repeated
over the years bringing the level of precision to 70 parts per million3.

3. References
1. R. V. Pound and G. A. Rebka, Jr. Resonant Absorption of the 14.4-kev
gamma
Ray from 0.10-μsec Fe57*Lyman Laboratory of Physics, Harvard University,
Cambridge, Massachusetts, Phys. Rev. Lett.* Received 23 November 1959
2. R. V. Pound and J. L. Snider, Effect of Gravity on Nuclear Resonance,
Phys. Rev.
Lett. 13, 539 (1964).
3. Tests of relativistic gravitation with a spaceborne hydrogen maser"
R.F.C. Vessot, M.W. Levine, E.M. Mattison, E.L. Blomberg, T.E. Hoffman,
G.U.
Nystrom, B.F. Farrell, R. Decher P.B. Eby, C.R. Baugher, J.W. Watts, D.L.
Teuber and F.D. Wills, Physical Review Letters, Vol. 45, Dec. 1980, pp.
2081-
2084

REFERENCES

1. Greene, Brian. The Elegant Universe. London: Vintage 2000.
2. Greene, Brian. The Fabric of the Cosmos. New York: Random Books, 2004.
3. Estate of Albert Einstein. Relativity. New York: Random House, 1961.
4. Einstein, Albert. The meaning of relativity. Princeton: Princeton University Press, 1988.
5. Hawking, Stephen. A Brief History of Time. New York: Bantam Books, 1988.
6. Hawking, Stephen. Black Holes and Baby Universes. Great Britain: Bantam Books, 1994.
7. Zukav, Gary. The Dancing Lu Masters. New York: Bantam Books, 1979.
8. Thuan Xuan, Trinh. The Changing Universe:Big Bang and After. New York: New Horizons, 1993.
9. Retic, H.E.The Einstein Hoax. New Jersey: 1st Books Library, 2001.
10. Tucker, Wallace and Allan. The Dark Matter. New York: Quill, 1988.
11. Feynmann, Richard. Six Easy Pieces. California: Perseus Books, 1995.
12. Feynmann, Richard. Six Not So Easy Pieces. California: Perseus Books, 1997.
13. Bolstein, Arthur. Ordinary Failure of an Extraordinary Theory. Atlanta: Protea Publishing, 2001.
14. Bodanis, David. E=mc2, A Biography of the World's Most Famous Equation. G.B: PanMacMillan, 2000.
15. Moore, Patrick. 2000 Year Book of Astronomy. London: MacMillan, 1999.
16. Dormand, John & Woolfson, Micheal. The Origin of The Solar System. Great Britain: John Wiley and Sons, 1989.

17. Graham-Smith, Francis & Lovell, Bernard. Pathways to the Universe. New York: Cambridge University Press, 1988.

18. Mather, John & Boslough, John. The Very First Light. Great Britain: Penguin Books, 1998.

19. Strathern, Paul. The Big Idea- Einstein and Relativity. New York: Anchor Books, 1999.

20. White, Micheal & Gribbin, John. Einstein: A life in Science. Great Britain: Free Press, 2005.

21. Kaku, Michio. Einstein's Cosmos. Great Britain: Phoenix Popular Science, 2005.

22. Kaku, Michio & Thompson, Jennifer. Beyond Einstein. New York: Anchor Books, 1995.

17. Graham-Smith, Francis & Lovell, Bernard. Pathways to the Universe. New York: Cambridge University Press, 1988.

18. Mather, John & Boslough, John. The Very First Light. Great Britain: Penguin Books, 1998.

19. Strathern, Paul. The Big Idea- Einstein and Relativity. New York: Anchor Books, 1999.

20. White, Micheal & Gribbin, John. Einstein: A life in Science. Great Britain: Free Press, 2005.

21. Kaku, Michio. Einstein's Cosmos. Great Britain: Phoenix Popular Science, 2005.

22. Kaku, Michio & Thompson, Jennifer. Beyond Einstein. New York: Anchor Books, 1995.

**TAREK SAMI AHMADIEH, B.E, Civil Engineering-
Graduate of the American University of Beirut, Lebanon,
has long found the topic of relativity to be as controversial as
the basis it is built on. He uses a logical and neat approach to
highlight the flaws of Relativity and to point at the numerous
paradoxes it creates. Relativity and the tell-tale properties of
Light has been one of his favorite subjects since his teenage
years. He enables the reader to look at these matters from a
new perspective, makes his explanations understandable to
most, at the same time being clear to all people conducting
researches in the fertile valley of Relativity. He thus brings
forward in this interesting book a hand full of new ideas and
state of the art solutions that allow us to peep into new possible
horizons in the matters of cosmology and theoretical physics.
He holds several design patents relating to electrical and
mechanical equipment.**